ISBN 978-1-334-30479-8
PIBN 10711325

_____PREDICTION OF THE EFFECTS OF

RESTORATION OF EL CAPITAN MORAINE,

YOSEMITE NATIONAL PARK

Gary M. Smillie

William L. Jackson

and Mike Martin

Technical Report NPS/NRWRD/NRTR-95/48

WATER RESOURCES DIVISION

National Park Service - Department of the Interior
Fort Collins - Denver - Washington

The National Park Service Water Resources Division is responsible for providing water resources management policy and guidelines, planning, technical assistance, training, and operational support to units of the National Park Service. Program areas include water rights, water resources planning, regulatory guidance and review, hydrology, water quality, watershed management, watershed studies, and aquatic ecology.

Technical Reports

The National Park Service disseminates the results of biological, physical, and social research through the Natural Resources Technical Report Series. Natural resources inventories and monitoring activities, scientific literature reviews, bibliographies, and proceedings of technical workshops and conferences are also disseminated through this series.

Copies of this report are available from the following:

National Park Service   (303) 225-3500
Water Resources Division
1201 Oak Ridge Dr. Suite 250
Fort Collins, CO 80525

Technical Information Center   (303) 969-2130
Denver Service Center
P.O. Box 25287
Denver, CO 80225-0287

PREDICTION OF THE EFFECTS OF RESTORATION OF

EL CAPITAN MORAINE, YOSEMITE NATIONAL PARK

Gary M. Smillie, Bill Jackson, and Mike Martin

Technical Report NPS/NRWRD/NRTR-48

March 1995

Water Resources Division
Fort Collins, CO 80201

United States Department of the Interior
National Park Service
Washington, D.C.

iv

TABLE OF CONTENTS

INTRODUCTION

In the late 1800s El Capitan Moraine was blasted and reduced in elevation to decrease flooding in Yosemite Valley. The moraine serves as a hydraulic control for the Merced River in the central chamber of the valley. This change in base level is suspected of causing headward incision of the channel. Along with reducing the flooding frequency, channel incision may have caused a general lowering of ground water allowing encroachment of the forest into meadow areas near the river. A 1978 San Francisco State University Master's thesis presents evidence of channel degradation of the Merced River and tributaries (Milestone 1978). In a follow-up proposal to the National Park Service, Milestone suggested restoring the historic elevation of the moraine in an attempt to recreate preexisting conditions (Milestone 1980). Since submittal of this proposal, the park has been interested in the project but park management has not had sufficient information to determine whether or not resource-condition objectives would be achieved. This report describes an investigation to predict the effects of restoring the historic elevation of El Capitan Moraine in Yosemite National Park (YOSE). We suggest that the proposed moraine reconstruction will have only modest effects on channel capacities and associated ground water tables, and will be unlikely to reestablish historic (pre-blast) morphologic conditions.

OBJECTIVES

The objective of this study is to evaluate the potential for restoring a more naturally functioning stream and riparian system in the central chamber of Yosemite Valley by reestablishing 4.5 feet (ft) of elevation to El Capitan Moraine. Three specific effects are investigated: 1) the extent and nature of sediment deposition upstream of the moraine to assess whether or not a more natural stream configuration would result, 2) the effect on flooding frequency in the valley due to sediment deposition in the channel, and 3) hypothesize the effect on groundwater elevations and associated potential impacts to meadow vegetation.

METHODS

The depth of sedimentation and the length of reach affected by restoring the historic elevation of El Capitan Moraine were estimated using a method developed from experience with small dams (Van Haveren, et al. 1987). This research shows that as sediment accumulates behind an impoundment, a new equilibrium slope is eventually established that is less than the original slope. This results in the formation of a "wedge" of deposited sediment, whereby the depth of aggraded sediment is greatest immediately upstream of the dam with ever smaller deposition upstream until at some point no aggradation occurs. In the above mentioned study, ratios of new slope to original slope were generally found to be in the range of 0.60 to 0.80. A ratio of 0.75 was used in this work to produce an "optimistic" estimate of potential channel aggradation in the Merced River.

Survey and hydraulic information for the Merced River and floodplain from El Capitan Moraine to approximately Stoneman Bridge was obtained from the U.S. Army Corps of Engineers (COE). The COE had collected channel and floodplain cross sections, distances between cross sections, and a measure of hydraulic roughness at each cross section (Manning's "n") in an earlier flood study of the area. For purposes of estimating sedimentation, the cross section data and distance between cross sections were used. First, the cross section located at the moraine was changed by assuming a horizontal channel bottom 4.5 ft higher than the thalweg (the moraine is estimated to have been lowered by about 4.5 ft). Next, a new longitudinal channel profile was estimated by reducing the original slope from cross section to cross section by 0.75. For example, if the change in elevation from one cross section to the next was one foot, pre-dam, then the change in elevation, post-dam, is assumed to be 0.75 ft. This was done in an upstream direction until the new bed profile intersected the original bed profile.

Water surface elevations for various flows were estimated using the COE computer hydraulics model, HEC-2. The model was run using both the existing channel profile and the assumed aggraded profile and the results were compared to identify the effect of the moraine on water surface elevations. The discharges used in this study were derived using log-Pearson III analysis of the U.S. Geological Survey gaging record, Merced River at Pohono Bridge. The period of record used in this analysis consisted of 67 years from 1917 to 1983. Magnitudes of flow associated with various recurrence intervals were computed and are shown on Figure 1.

RESULTS AND DISCUSSION

The Merced River has a very low gradient in the central chamber of Yosemite Valley, approximately 0.12% or about 6.25 ft per mile. This low gradient causes sedimentation to be experienced for a considerable distance upstream due to an increase in base level downstream. According to the technique used, sediment will accumulate in the Merced River channel for a distance of about 20,600 ft upstream from the moraine site. The depth of this sediment will be 4.5 ft at the moraine site and will decrease upstream to zero at the upper extent (Figure 2). An undetermined but relatively long period of time would be necessary to attain a new equilibrium slope.

Water surface elevations are predicted to be higher for a given flow with the aggraded channel than in the existing condition. The magnitude of the increase, however, is less than the increase in channel elevation because of greater width of flow at higher stages (Figure 2). To asses the impact of moraine reconstruction on flood frequency, the flows required to just fill the channel were calculated for three representative cross sections within the impacted reach (Table 1).

At the lowest cross section (through the lower portion of El Capitan Meadow) a flow with a recurrence interval of approximately 200 years is necessary to reach the top of bank with the existing channel. This is consistent with a claim by YOSE staff that flooding in El

Capitan Meadow is very rare. A recurrence interval of 200 years for bankfull flow is unusually high, but can be explained by the deeply incised channel present at this location. Bankfull flow in the aggraded channel at the same location is approximately the 100-year flow. This recurrence interval for bankfull flow is still unusually large and an explanation is not obvious. At the second cross section (about midway up the impacted reach) approximately, the 10- and 5-year flows are required to just fill the channel in the pre- and post-reconstruction settings, respectively. At the third and highest cross section (located near Yosemite lodge) about a 1.4-yr flow will fill the channel before sedimentation, and about a 1.25-yr flow will fill the channel after sedimentation. Interestingly, bankfull flow has been shown to commonly occur with a recurrence interval of approximately 1.5 years suggesting this area of the stream has been largely unaffected by the reduction in base level at the moraine.

Table 1. Characteristics of Merced River before and after restoration of elevation of El Capitan Moraine. Two cases of restoration are shown: 4.5 feet and nine feet of added elevation.

	No Modification	4.5 feet	9.0 feet
Length of Sedimentation (Feet)	0	approx. 20600	approx. 22900
Recurrence Interval of Bankfull Flow (Years)			
XS #4	200	100	~1.7
XS #15	10	5	~1.7
XS #23	1.4	1.25	~1.0

To evaluate the potential impact to groundwater elevations, it is assumed that the Merced River is underlain by an unconfined alluvial aquifer bounded on each side of the valley by impermeable rock. The effect of channel aggradation on ground water would be to raise the water table to the new river water surface elevation (Figure 3-5). This effect would be greatest at the surface water/aquifer boundary and would diminish with distance from the river, and with distance upstream from the moraine. The greatest increase in water table elevation would be less than 4.5 ft close to the river banks, just above the moraine (Figure 3). At greater distances laterally from the river and upstream from the moraine, water table increase will be significantly less (Figure 4 and 5). Additional study is needed to determine the characteristics of meadows with stable vegetation, and at that time an evaluation of any impacts to non-meadow vegetation could be made. However, given the small increase in groundwater elevation over most of the area, it is unlikely that encroachment of non-riparian vegetation on the floodplains will be significantly reduced except possibly in the immediate vicinity of the moraine.

As mentioned above, a very large recurrence interval is predicted for bankfull flow near the moraine site, about 100 years. One possible explanation is that the moraine may have been reduced in elevation more than the four to five feet estimated by Milestone. To test this hypothesis, we made an additional run with the hydraulics model using an assumed base level change of nine feet. The analysis was done in an identical manner to the previous analysis. With the assumed nine feet of restoration, the length of reach predicted to aggrade would be approximately 22,900 ft. The frequency of bankfull flow for both cases are summarized in Table 1. As can be seen in the table, the frequency of bankfull flow near the moraine has changed dramatically, from about a 100- to about a 1.7-year (yr) recurrence interval. At the intermediate cross section the frequency has changed from about a 5- to about a 1.7-yr recurrence. The frequency of bankfull flow at the highest cross section changed very little. All of these newly calculated recurrence intervals are in the range of usual natural recurrence intervals for bankfull flow. While this evaluation is by no means definitive, it does suggest that perhaps more material was removed from the moraine than originally thought.

CONCLUSIONS AND DISCUSSION

By analyzing existing channel capacities, this investigation adds support to the hypothesis that the 19th century moraine blasting resulted in down-cutting of the Merced River in the central chamber of Yosemite Valley (Milestone 1978). The modeling analysis clearly portrays a river with unusually large capacity in the vicinity of the moraine, and considerably smaller capacity upstream near Yosemite Lodge. This increased channel capacity immediately upstream from the moraine is consistent with the suggestion that the channel has experienced down-cutting. However, our investigation also suggests that while some channel aggradation would result from moraine reconstruction, it is unlikely that pre-moraine blast river conditions would be reestablished. The central chamber of Yosemite Valley was formed by lacustrine processes after glacial retreat which were subsequently reworked by river flows. These processes led to a wide, and exceptionally low gradient floodplain and channel. Channel shifting and flooding were common under these circumstances. Reduction in base level has caused incision of the channel with greater bank stability and less frequent flooding. The proposed restoration would take place in a fluvial environment and replace sediment in a different manner with only some of the characteristics of the natural system.

Specifically, moraine reconstruction would result in modest increases in flooding, moderate (0 - 4 ft) increases in riparian water tables, and possibly a slight increase in lateral channel adjustment processes. All of these effects would decrease in magnitude with increased distance upstream from the moraine. Additionally, the results of this study indicate that the moraine may have been reduced in elevation more than the four to five feet estimated in earlier studies. However, more study is needed to substantiate this speculation.

REFERENCES

Milestone, J.F. 1980. A study examining the possibilities of restoring the original base-level of erosion for the central chamber of Yosemite Valley, California. Unpublished.

Milestone, J.F. 1978. The influence of modern man on the stream system of Yosemite Valley. MS thesis, San Francisco State University.

Van Haveren, B.P., W.L. Jackson, and G.C. Lusby. 1987. Sediment deposition behind Sheep Creek Barrier Dam, southern Utah. *Journal of Hydrology* 26(2).

6

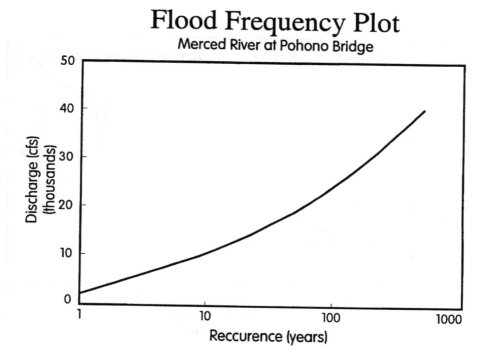

Figure 1. Discharge versus recurrence interval for the Merced River at Pohono Bridge.

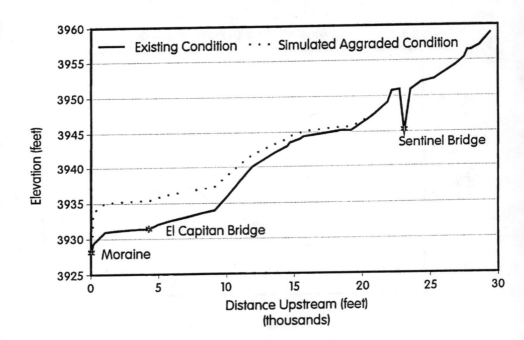

Figure 2. Original bed profile and aggraded bed profile for the Merced River.

8

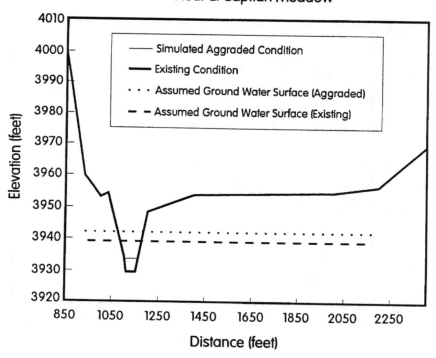

Figure 3. Cross section No. 4, Merced River near El Capitan Meadow.

9

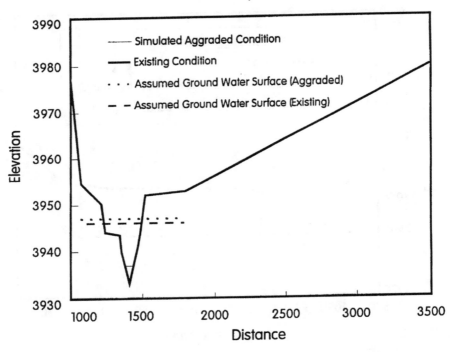

Figure 4. Cross section No. 15, Merced River above El Capitan Bridge.

10

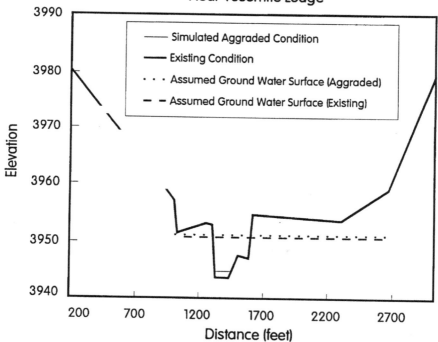

Figure 5. Cross section No. 23, Merced River near Yosemite Lodge.

As the nation's principal conservation agency, the Department of the Interior has the responsibility for most of our nationally owned public lands and natural and cultural resources. This includes fostering wise use of our land and water resources, protecting our fish and wildlife, preserving the environmental and cultural values of our national parks and historical places, and providing for enjoyment of life through outdoor recreation. The Department assesses our energy and mineral resources and works to ensure that their development is in the best interests of all our people. The Department also promotes the goals of the Take Pride in America campaign by encouraging stewardship and citizen responsibility for the public lands and promoting citizen participation in their care. The Department also has a major responsibility for American Indian reservation communities and for people who live in island territories under U.S. administration.

NPS D-491   March 1995

CPSIA information can be obtained
at www.ICGtesting.com
Printed in the USA
BVHW071018141218
535632BV00019B/895/P